# DAY AND NIGHT

## AS THE EARTH TURNS

Lynn M. Stone

The Rourke Book Co., Inc.
Vero Beach, Florida 32964

Edited by Sandra A. Robinson

PHOTO CREDITS
Courtesy NASA: cover; © Lynn M. Stone: title page, pages 4, 8, 10, 12, 13, 15, 17, 18, 21; © Hal Jandorf: page 7

**Library of Congress Cataloging-in-Publication Data**

Stone, Lynn M.
Day and night / by Lynn M. Stone.
p. cm. — (As the earth turns)
Includes index.
ISBN 1-55916-022-5
1. Day—Juvenile literature. 2. Night—Juvenile literature.
3. Earth—Rotation—Juvenile literature. [1. Day. 2. Night.
3. Earth—Rotation.] I. Title. II. Series: Stone, Lynn M.
As the earth turns.
QB633.S75 1994
525'.35—dc20 93-39664
CIP
AC

# TABLE OF CONTENTS

## DAY AND NIGHT

You can wish for 24-hour days of sunshine, but they will never happen. Each 24-hour day has a period of light and a period of darkness, too. The movement of the Earth in space guarantees it.

To understand how night and day happen, start with a basketball. The basketball is shaped like Earth. Now spin the ball ...

*Sunrise brings daylight after a period of darkness*

## SPINNING IN SPACE

Imagine the basketball is Earth and you are the sun. During each spin, the basketball label faces you and then turns away from you.

The Earth spins as it travels in **orbit,** a wide circle, around the sun. The spinning motion gives us a daily dose of darkness and of light.



*While one side of the Earth faces the sun and has daylight, the other side has night*

## THE SUN'S LIGHT

Our natural light on Earth comes from the sun. The sun is a huge ball of fire.

The sun is so powerful that some of its heat and light reach 93 million miles through space to Earth.

When our part of the Earth faces the sun, we have daylight. As we spin away from the sun, we move into night.

*The sun's energy reaches 93 million miles through space to heat and light the Earth*

OCEAN
ATLANTIC OCEAN

## ORBITING THE SUN

Remember that basketball? Hold the ball at eye level and look at the label. The basketball has a top and bottom. The Earth has a top and bottom, too. They are called the north and south poles.

Each year the Earth makes one orbit around the sun. During this journey, the Earth spins at a slight tilt, or angle, like a leaning top. Because of that, each pole takes a turn tilting more toward the sun during the year.

*Studying a globe is a good way to see the tilt of the Earth's poles*

*As a pelican's part of the world spins away from the setting sun,*
*the pelican's workday ends*

*Twilight is in the eastern sky just before the sun rises above the horizon*

## CHANGING DAYS

We in North America live in the northern **hemisphere,** the northern half of the Earth. We are much closer to the north pole than the south pole.

When the north pole tilts toward the sun, we enjoy spring and summer. The sun is then striking the northern half of the Earth more directly than the southern half.



*The northern hemisphere enjoys summer while the north pole is tilted slightly toward the sun*

## SHORT AND LONG DAYS

The day with the longest period of sunlight in the northern hemisphere is on or about June 21. On that same day, because of the Earth's tilt, the southern hemisphere has its shortest day.

Australia and other countries in the world's southern half have their longest day on December 21. That is when the northern half of the world has its shortest period of sunlight.



*Midnight in the Far North, where summer days are so long it's called the "Land of the Midnight Sun"*

# DAYLIGHT

The sun appears in the sky during daylight hours. Daylight makes most of the things we do easier. Most outdoor work and play takes place in daylight. Schools and businesses are open mostly during daylight hours.

Daylight is even more important because plants change daylight's sunshine into food.

The sun produces heat as well as light. Daytime, with its daylight, is warmer than night.

*By providing plants with food energy, the sun is the basic source of life on Earth*

## NIGHT

As one part of the Earth spins away from the sun, it loses sight of the sun. The sun sets, we say, and darkness follows. It is nighttime. Many businesses close until morning, but many animals are **nocturnal** — active at night.

Most of our nighttime light is produced by electricity. Sometimes sunlight is reflected off the moon toward the Earth. Then the moon is shining.



*Moonlight washes the Atlantic Ocean, as night follows day*

## TWILIGHT

Each morning the sun rises in the East. It may look quite close, but it is still 93 million miles away. Each evening the sun sets in the West. We can't always view sunrises and sunsets, however, because clouds may hide them.

Just before sunrise and just after sunset, the sun's light looks soft. The sky may be very colorful. These times of day are called **twilight.**

## Glossary

**hemisphere** (HEHM iss fear) — either the northern or southern half of the Earth, using the equator as a divider

**nocturnal** (nahk TUR nul) — active at night

**orbit** (OR bit) — the path that an object follows as it repeatedly travels around another object in space

**twilight** (TWI lite) — the period of low light just after sunset and just before sunrise

# INDEX